IMAGES
*of America*

# STODDARD

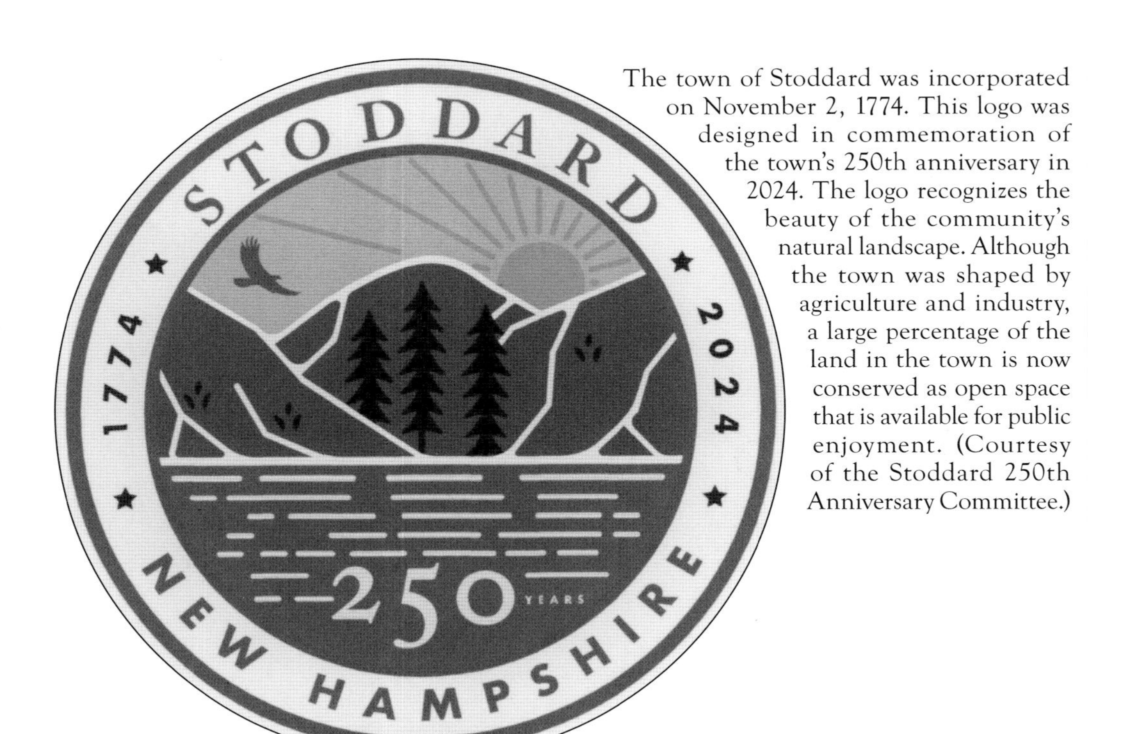


The town of Stoddard was incorporated on November 2, 1774. This logo was designed in commemoration of the town's 250th anniversary in 2024. The logo recognizes the beauty of the community's natural landscape. Although the town was shaped by agriculture and industry, a large percentage of the land in the town is now conserved as open space that is available for public enjoyment. (Courtesy of the Stoddard 250th Anniversary Committee.)

**On the Cover:** The steamboat *Myra* was photographed by Charles Newell in the late 1800s. The *Myra*'s presence on Highland Lake illustrated an important transition occurring in Stoddard. As an industrial tool, the steamboat hauled logs for the Stoddard Lumber Company, but more importantly, it also transported visitors across the large lake who were among the new wave of "back-to-nature" tourists flocking to the town at the turn of the 20th century. (Courtesy of the Stoddard Historical Society.)

IMAGES
*of America*

# STODDARD

Alan F. Rumrill

ISBN 978-1-4671-6061-2

Published by Arcadia Publishing
Charleston, South Carolina

Printed in the United States of America

Library of Congress Control Number: 2023940680

For all general information, please contact Arcadia Publishing:
Telephone 843-853-2070
Fax 843-853-0044
E-mail sales@arcadiapublishing.com

Visit us on the Internet at www.arcadiapublishing.com

# Contents

# Acknowledgments

Charles L. Peirce first arrived in Stoddard by bicycle in the summer of 1893 from his home in Beverly, Massachusetts. He came to visit a schoolmate and was immediately intrigued by the history of the quiet community and the personalities of the friendly, welcoming residents. He purchased an old farmhouse in Stoddard less than 10 years later and served as unofficial town historian for the next half-century.

Peirce compiled and preserved extensive historical collections about the town. An essential part of his work involved recording many scenes in the community on film so that rapid changes occurring there would be preserved visually to enhance the written documents he was collecting and creating. This book of photographic history would have been impossible to write without the lifelong dedication of Stoddard historian Charles Peirce.

Another amateur photographer also added considerably to this story. Charles Newell used his interest in photography to record important changes in the community during the late 1800s and early 1900s. Newell was a Boston banker who summered in Stoddard with his family beginning in 1889. His dozens of detailed photographs illustrated the town as it transitioned to a summer retreat community.

Several collectors of Stoddard photographs also played crucial roles in the compilation of this work. The Stoddard Historical Society and the Historical Society of Cheshire County have both collected historical photographs of the town for almost a century. They generously allowed the use of their images in this volume.

The accumulations gathered by three private collectors also added to this work. My grandmother and Stoddard historian Pheroba Wilson collected Stoddard photographs for decades. Mike Eaton and David Sysyn also compiled large collections of Stoddard images because of family connections to the town. The hundreds of photographs collected by these three people enhanced this book.

I also wish to thank the amateur photographers who have been inspired to record the town's natural beauty and historic character on film over the past 160 years. Finally, thanks to my wife, Kim, who encouraged me and allowed me time to complete this work during a very busy time in our lives.

In addition to the Stoddard Historical Society (SHS) and the Historical Society of Cheshire County (HSCC), many of the photographs in this book are from the collection of the author. Those photographs are identified as (COTA) in the courtesy line.

# Introduction

The town of Stoddard, New Hampshire, located in the southwest corner of the state and near the geographic center of New England, celebrates the 250th anniversary of its incorporation in 2024. Since its settlement by farmers of European descent in 1768, the town has undergone three major transitions in its development that have resulted in the preserved historical character and expansive natural beauty for which it is known today. Like many hill towns of central New England, the town has experienced successive periods when it relied on agriculture, industry, and tourism for financial support.

The settlers who began to arrive in 1768 relied almost exclusively on agriculture for their livelihood. The population grew quickly, from zero White residents in 1767 to more than 200 when the Revolutionary War began less than a decade later. Hundreds of farm families arrived, mostly from Massachusetts and Connecticut towns to the south. For three generations, they cleared the land, grew crops and livestock, and in the early 1800s, added thousands of sheep to their farms to supply wool to local families and to the mills of New England's burgeoning woolen industry.

Stoddard reached a peak population of 1,203 people in 1820. By that time, however, the town's farms were beginning a long, slow decline that would continue for 100 years. Farm families faced depletion of the thin, rocky topsoil and competition for the sale of cash crops from the richer soil being tilled to the west. Stoddard farmers and their families left town to seek some of that western land or to work in the mills of New England's growing industrial towns and cities. Several dozen abandoned farmhouses dotted the landscape by the 1870s.

Industry grew in importance as agriculture declined. Glass manufacturing was introduced in the early 1840s and helped to support the declining population for a generation in the mid-1800s. Five glass factories produced millions of bottles for markets across the Northeast between 1842 and 1873 and provided employment for hundreds of residents.

Numerous woodenware factories also operated during the 19th and early 20th centuries. Mills manufactured boxes, pails, ox yokes, chairs, powder kegs, rakes, rolling pins, baseball bats, lumber for building, and numerous other products. The factories made use of the trees being cleared for the development of farms and then made use of the white pine trees that grew on abandoned farms after the middle of the 19th century. The Stoddard Lumber Company purchased thousands of acres of former farmland during the last half of the 1800s to supply wood for this firm, which grew to become one of the largest businesses in the history of the town.

Despite temporary success, the factories began to disappear by the late 1800s. Several economic factors hastened this decline, chief among them being the cost and difficulty of transporting finished products to distant markets. The railroad was never constructed through Stoddard, and the mills in the community could not afford the added financial burden of transporting their goods to the railroad. The extensive facilities of the Stoddard Lumber Company burned to the ground in 1908, terminating a business that was already in decline.

The population of the town dwindled to 257 residents in 1910, a total of 213 in 1920, and just 113 people by 1930. There was some question about whether the town would survive. However, a new national pastime and a new technological development combined to revive the community.

By the last decades of the 1800s, Americans with spare time and sufficient income were escaping to the countryside to enjoy nature. Authors and artists had been touting the peacefulness and beauty of nature since the mid-1800s, and Americans began to listen. By the early 1900s, national parks had been formed, portions of nearby Mount Monadnock were conserved, and outdoorsman and Pres. Teddy Roosevelt encouraged the preservation of natural space for the use and enjoyment of all Americans.

Stoddard was a natural destination for the increasing number of people who were traveling from urban areas to take advantage of the physical and psychological benefits provided by nature. The town offered nine lakes and ponds and large tracts of former farmland that were now forested hillsides and valleys. These open spaces were ideal for hunting, fishing, camping, swimming, boating, hiking, and simply enjoying the beauty of nature.

Adding to the enjoyment for some visitors was the somewhat arduous nature of the journey to out-of-the-way communities like Stoddard. They traveled by train and then by stage, by wagon, or on foot to approach these isolated towns. That all changed with the introduction of the automobile, however. By the 1920s, many middle-class families could afford to buy a family car and drive over the old roads to towns like Stoddard.

The number of summer visitors increased through the early years of the 1900s. Those visitors soon began to buy land and build summer homes along the shores of the lakes and ponds, becoming summer residents rather than temporary tourists. The population of permanent residents increased slowly, but the summer residents doubled the population during the months of June, July, and August. Furthermore, improvements in automotive technology meant that cars became reliable daily transportation to and from jobs some distance away. The town became a bedroom community for those who wished to reside where they were surrounded by nature but earn a good wage working at jobs some distance away.

Stoddard survived the population nadir of 1930. The number of residents has increased dramatically since 1960, when the population numbered just 146 people. Most residents continue to work outside of town. The lakes are ringed with hundreds of year-round and summer homes. However, the lack of industrial and commercial development for a century meant that the historic villages have changed very little since the mid-1800s. They offer the type of authentic, unchanged history that attracts heritage tourists who long to learn about life in small-town New England in the past.

The wooded hillsides and numerous bodies of water crowded between them are still magnets for outdoor enthusiasts as well. The introduction of conservation easements and the increase in land conservation organizations have had a major impact on the evolution of the town over the past 50 years. Many abandoned farms were acquired by individuals or families, creating large tracts of forest across the community. Many local landowners wished to preserve their property as open natural land for the enjoyment of future generations. Approximately two-thirds of the land in Stoddard is now preserved by the town, by conservation organizations, or through conservation easements and will not be developed. Most of that land is open for public enjoyment, attracting residents and outdoor enthusiasts alike through New England's four seasons.

Stoddard, New Hampshire, survived the lean years following the demise of agriculture and industry to reemerge as a thriving oasis of natural beauty and New England heritage. After almost 200 years, the population of the small town once again surpassed 1,200 residents. All these struggles and successes were preserved on film following the introduction of photography in the mid-1800s. It is hoped that the images shared here will enrich the story of this unique community.

## *One*

# STODDARD CENTER

Stoddard Center, the third center village in the town, was developed in the 1830s when businesses and institutional buildings were constructed near the intersection of four roads. This image from 1867 or 1868 shows farmers harvesting hay in the foreground adjacent to the Congregational church and its carriage sheds. The hills in the distance had been cleared for agricultural purposes. (HSCC.)

The Stoddard Congregational Church, built in 1836, and the district No. 1 elementary school, built in 1850, stood side-by-side for 129 years at the west end of the village. This image shows the buildings in about 1920, after the soldiers' monument was installed on the church lawn. This was eventually the only school in town as the population of the community decreased throughout the 19th and early 20th centuries. (COTA.)

Snow filled the village main street in March 1926 when historian Charles Peirce ventured out on snowshoes to record the appearance of his adopted community. The church appears in the distance, and the Stoddard Historical Society building is at the right. A telephone pole appears on the left side of the street in the foreground, but it would be 13 more years before electricity came to town. (COTA.)

This view shows the village center on a summer afternoon in about 1920. The Central House hotel, on the left, was a popular hostelry here for a century. This bucolic view at the intersection of Forest Road, Center Pond Road, and Mount Stoddard Road shows a boy strolling in the middle of the quiet street, a feat that is often possible on the lightly traveled road even today. (HSCC.)

Historic homes lined the gravel main street, known as Forest Road, in the 1920s. The house at the far left was built by storekeeper George Ireland in 1843. The Stoddard Congregational Society purchased the house in 1861, and it served as the town's parsonage for several decades. The Congregational minister, Samuel Gerould, joined the Army during the summer of 1862 to aid the Union cause during the Civil War. (COTA.)

The Morse house, often referred to locally as the Morse mansion, was built in 1843. Town cobbler Nathan Morse bought the old town meetinghouse at auction in June 1843 for a bid of $105.37½. He built his large residence in the village later that year using the timbers and other elements of the previous meetinghouse, incorporating benches and other features of the building in his new home, above. The house was in a state of disrepair by 1970. The Morse house was rescued from collapse by Robert and Joanne Webster, who restored the house and returned it to its previous glory by the time of the town's bicentennial celebration in 1974, as seen below. (Above, COTA; below, HSCC.)

There was no traffic on the village street on this summer day around 1920. The town hall, visible to the left, was built in 1868 and still serves that purpose today. The road that branches to the left in front of the town hall joined King Street. Most of the farms located along King Street were abandoned and reclaimed by forest by the time this photograph was taken. (HSCC.)

The Nathan Gould house, shown here in the 1890s, is located adjacent to the town hall. This house was originally erected about two miles from the village. Gould, a farmer, manufacturer of ox yokes, and renowned cabinetmaker, moved his house to this location in the early 1830s. The former home now serves as Stoddard town government offices. (COTA.)

This winter scene was taken from the apple orchard behind the church on the last day of 1901. Two general stores and the Central House hotel are visible in this view. Today there are no retail enterprises in the village. The population of the town decreased by more than two-thirds between 1820 and 1900 as farm families left the failing hill farms and the last glass factory closed in 1873. The number of barns and other outbuildings in the village, and the extensive open pastures visible on the distant snow-covered hills, illustrate the agricultural heritage of the community. Those barns are all gone today, and the open land is now completely reforested. Thousands of acres of land on the hills in the distance have now been protected by conservation easements and are open to the public for recreational purposes. (COTA.)

A solitary car is rolling down Forest Road in this early-20th-century postcard image. Horseshoe tracks are visible in the road, as the population transitioned slowly to the new automobile transportation. Ancient maples line the street, and the new flagpole, installed in front of the town hall during World War I, is visible in the distance. (COTA.)

Stearns Foster built his home in 1840. This 1925 view was taken soon before the house was acquired by architect Howard Goodspeed. He removed the barns and remodeled and enlarged the house. The Goodspeed family lived there for several decades before the house was converted for use as the Pitcher Mountain Inn in 1978. (COTA.)

This c. 1940s view captures the New England character of the historic villages of the southwest corner of the state, which is referred to as the "Currier and Ives" corner of New Hampshire. The Congregational church, Central House hotel, and the Hadley house, all built between 1833 and 1836, are arranged from left to right in this image. (HSCC.)

Isaac Duncan completed his home in the village in 1843. Following the death of Isaac and his wife, Rebecca, the home was used as a summer boardinghouse by Benjamin H. Griffiths. The attached two-story barn was remodeled for use as guest rooms. The building was being used as a grocery store when it burned in 1911. The Davis Public Library is now located on this site. (SHS.)

# *Two*

# AGRICULTURE

Stoddard was a farming community from its earliest settlement. Almost all the families that arrived between the 1760s and the 1820s tilled the rocky, thin soil on the hillsides. The farms were successful for 75 years, after which the farm families began to leave town for better opportunities elsewhere. The McClure farm, shown here on a cloudy winter day in the early 1900s, was typical of the hundreds of small farmsteads that dotted the landscape. (COTA.)

The Cold Spring Pond neighborhood was photographed in 1861. All the land had been cleared by diligent farmers, but they could not remove the countless rocks visible in this view. Phillip Abbott's small blacksmith shop, left of center, and Luther Abbott's carding mill, left, were important commercial operations in the farming community. The mill's water-powered carding machines relieved housewives of the tedious chore of carding wool by hand. (HSCC.)

This image of the Eben Stacy farm on Shedd Hill Road was taken in June 1888. This neat farmstead had been settled decades earlier in a neighborhood of Stacy farms east of what is now Highland Lake. Stacy Pond is visible behind the house. This small pond and two others had been joined, forming Highland Lake, when a dam was installed in Mill Village almost 20 years earlier. (SHS.)

The English-style barn at the Dodge farm near Center Pond was probably built in the 1820s. Later owner John T.W. Mountford is shown working in his garden in the early 1900s. This property was farmed into the 1940s, but the barn and accompanying farmhouse are now gone. The entire farm is completely reforested today. (COTA.)

Levi P. Spalding poses with his prized horse in the early 1900s. Horses were essential for transportation and farm labor before the introduction of automobiles and tractors. Spalding operated a tavern in the village of South Stoddard that catered to sheep and cattle drovers who stopped off for the night as they traveled to and from the pastures of southwest New Hampshire. (COTA.)

The small one-and-a-half-story portion of the Reed farmhouse near Center Pond was the original residence. As the farm and family grew, the large addition to the left was added. The outbuildings housed livestock, crops, and farm equipment. Notice the stone walls that were built to mark boundaries, enclose pastures and gardens, and remove rocks from land. (COTA.)

Harriet Spaulding poses with her shepherd's staff and her father Henry's sheep on Knowlton Hill in August 1913. Sheep herding became an important agricultural enterprise during the first half of the 1800s. The demand for wool grew with the rapid expansion of the textile industry in New England, and farmers cleared vast tracts of land to pasture sheep. (COTA.)

Henry Spaulding's cattle and horses mingle in their pasture on Knowlton Hill during the summer of 1913. By this time, most of the farms in the town had been abandoned. The few farmers who stayed in the town used the remaining pasturelands to graze their animals during the summer months. (COTA.)

The large barn on the Abner Knowlton farm is clearly visible at the center of this early 1900s photograph. This farm is located high on the slopes of Pitcher Mountain, the third-highest peak in southwest New Hampshire. Grazing sheep are visible in the foreground, and the region's namesake mountain, Monadnock, is barely visible in the distant summer haze. (COTA.)

The image above, from April 1902, shows Levi P. Spalding's cattle drive passing through Stoddard Center on its way to summer pasture in the town. Most of the farms were abandoned by the end of the 1800s. A few farmers who stayed became cattle dealers, pasturing their herds in the fields of Stoddard through the summer and then driving them to sell at the cattle market at Brighton, Massachusetts, in the fall. These drovers acquired farmland inexpensively, often by paying unpaid taxes on the abandoned property. This allowed some families to remain in the rapidly depopulating community. Harry Wilson's cattle await the end-of-summer drive to Brighton in his corral near the shore of Granite Lake below. (Both, COTA.)

Perley Swett is pictured with his herd of goats at his family homestead in the southwest corner of the town. Perley lived alone in the old homestead for decades, becoming known as "the Hermit of Taylor Pond." When other farm families left their homes, Perley was left alone in the woods. He once mused that it was easy to become a famous hermit simply by not moving away or dying. (COTA.)

Pitcher Mountain Farm has long been known for its herd of Scottish Highland cattle. The Faulkner family operated the 220-acre property as a Scottish Highland beef operation for half a century. Brian and Keira Farmer have operated the farm since 2013, when they brought their Scottish Highland cattle to the property. It is the only remaining commercial cattle operation in the town. (SHS.)

Millard Edwards managed the Pitcher Mountain Farm in the mid-1900s. He is shown here with two of the farm's Scottish Highland cattle and two young ladies dressed in their Scottish outfits. There has been a farm located at this site high on Pitcher Mountain since the 1780s. Nathaniel Emerson also operated a tavern from his home here in the 1790s. He was succeeded by Aaron Matson, the only Stoddard resident ever elected to the US Congress. Abner Knowlton owned the farm when the buildings burned in the 1860s. Farmer Henry Spaulding completed the reconstruction. The farm was known for a time as Pinnacle Stock Farm. It has been known as Pitcher Mountain Farm for more than half a century. The Faulkner family introduced Scottish Highland cattle to the property in the mid-1900s and developed one of the largest and best-known Scottish Highland stock farms in the East. (HSCC.)

*Three*

# Business and Industry

Joel Eaton built this large store in the new center village in 1840. The store contained the post office, a meat market in the basement, and offices for rent upstairs. Frank Reed bought the business in the late 1800s. Reed's store was the last in Stoddard Center when it closed due to declining business in the early 1900s. The structure was condemned and demolished in 1957. (SHS.)

William Stewart built his general store in Mill Village in the 1920s on the site of the Island House hotel, which burned in 1917. Although there was another store in the village, having the post office in this building brought additional traffic to the store. The bandstand in the tree out front elevated band members during concerts. (SHS.)

Charles and Mary Lou Eaton purchased Stewart's store in 1951. Charlie was photographed on the front porch that year with his three oldest sons: Charles, Michael, and Thomas. As Eaton's General Store, it offered gasoline, groceries, and general supplies and continued to house the town post office. The Eaton family operated the store for about half a century. (SHS.)

Mary Lou and Charlie Eaton pose in their store in the 1950s. It was a true general store where the growing number of summer residents congregating around Stoddard's lakes and ponds could find almost anything they needed during their temporary visits to the community. Mary Lou served as Stoddard's postmaster for many years. The post office boxes are visible at the left. (SHS.)

Highland Lake Variety Store was built in the spring of 1966 and opened for the summer season. The seasonal business was open during the summer months and on weekends during the fall to cater to cottage owners on Highland Lake. The store owner was pleased that the shop had experienced good luck and "much encouragement" from the customers. (COTA.)

Stoddard Gas & Grocery was photographed in 1995. This store on busy Route 9 is one of the few retail establishments between Keene and Hillsborough—a distance of 30 miles. Note the phone booth in front of the store and the gasoline price of $1.12 per gallon. This store is now known as Mr. Mike's. (SHS.)

The Mill Village General Store is located on the site of several previous retail establishments. Myra Robb ran a small store here in the early 1900s. A gas station was located on this site in the mid-1900s. Later, a hardware store was operated in the small left wing. The large wing to the right is a more recent addition. This photograph was taken in 1995. (SHS.)

The commodious Green Mountain House, better known as the Box Tavern, was built by John Robb in about 1833. This South Stoddard hotel was constructed at the junction of two roads that saw many travelers as they journeyed to and from Keene, Concord, Hancock, and Stoddard Center. The introduction of the glass industry to the village in 1842 added to the hostelry's business. It was destroyed by fire in 1888. (COTA.)

The introduction of the automobile changed the nature of life in the quiet town of Stoddard. Families could now travel farther, and they had to stop for gasoline and food. The Aw Kom' Inn filling station, located alongside Route 9 in South Stoddard, offered gasoline and lunch for travelers. Proprietor C.E. Guillow also rented boats for those who wished to fish or row at the nearby Robb Reservoir. (COTA.)

The one-pump Aw Kom' Inn filling station was succeeded by Nick's Place, which boasted two gas pumps. Nick's also served lunch and had a dance hall. Nick's Dance Pavilion was a busy place on the weekend. This was especially true during the 1920s and early 1930s, America's Prohibition years, when liquor was readily available from jugs stored under the floor of the pavilion. (COTA.)

An old farmhouse near Granite Lake was enlarged and a public dining room was added in the early 20th century. The dining room served as the Minniowa Tea House. Tea houses were popular as roadside restaurants during the early years of automobile travel. The location of the farmhouse alongside Franklin Peirce Highway, the main highway between Keene and Concord, made it a logical location for a tea house. (COTA.)

The Spalding Tavern in South Stoddard village was constructed in the 1850s as a boardinghouse for employees of the adjacent South Stoddard Glass Manufacturing Company. It later became a tavern and the home of the South Stoddard Post Office. Levi P. Spalding purchased the tavern in 1902, shortly before the photograph above was taken. The tavern was a popular stop for drovers, who could place their sheep or cattle in the adjacent pasture while they spent the night in the tavern. The photograph below was taken inside the tavern dining room during the first decade of the 1900s. The inn was converted to a residence by the 1930s. The building burned on New Year's Eve 1952. (Both, COTA.)

The Saturday stage from Hancock to Marlow makes its regular stop at the Central House in Stoddard Center on July 7, 1906. The full stage, above, is transporting summer visitors. The hotel was built by Jonathan Sanderson in 1833. Isaac Duncan was the first proprietor. The hotel remained a town landmark for more than a century. Stephen and Cornelia Harrington operated the hostelry for many years beginning in 1878. Cornelia took in guests well into the 20th century. Plans to renovate the structure in the 1940s did not come to fruition, and it was razed in 1952. Gertrude Thomas is shown inside the hotel during demolition below. The bar, benches, and woodwork from the taproom were incorporated into the Spouter Tavern at Mystic Seaport Museum. (Both, SHS.)

The Island House in Mill Village was Stoddard's first summer resort hotel. The former residence was converted into a hotel by Rodney Brown in 1876. A dam that had been built on a brook behind the building a few years earlier created a long lake that drew outdoor enthusiasts to the village. The Island House, shown above in the 1880s, enticed potential visitors with the promise of unrivaled boating and fishing and the unspoiled beauty of nature. The hotel was famous for its hornpout stew, made with fish pulled from the lake out back. The picture below shows the front of the Island House around 1908. The hotel burned to the ground at 3:00 on the morning of March 14, 1917. The final two boarders escaped the flames by jumping out a window in their night clothes. (Above, SHS; below, COTA.)

Glass manufacturing, an important new industry and employment, began in the town in 1842 just as hill farms were being abandoned by the dozen. This is the only known photograph of the interior of any of the five glass factories. The New Granite Glassworks, shown here, opened in 1861. The five factories employed hundreds of residents and produced millions of bottles for Northeastern markets between 1842 and 1873. (HSCC.)

Gristmills such as this one in Mill Village were essential to grind grain for local farm families. Silas Wright began this mill in the 1700s. This and later water-powered mills in the immediate vicinity gave the village its name. This was the last gristmill in town, operating until about 1879. (SHS.)

Carding mills were an important development in the manufacture of cloth. These mills alleviated the necessity of time-consuming hand carding of wool before making it into cloth. Luther Abbott opened this carding mill in 1854 near the outlet of Cold Spring Pond. Abbott and other owners operated the mill until about 1900. The building still stands today. (SHS.)

The woodenware industry was the town's most important business. Wood and wooden products supplied jobs and income for thousands of residents. These logs were being floated down Long Pond, now Highland Lake, to the mills of the Stoddard Lumber Company, probably the largest individual firm in the history of the community. (SHS.)

Many logging crews and companies set up operations beginning in the late 1800s after the abandoned farmlands grew back to harvestable timber. This early-1900s logging camp was located near the village of South Stoddard. These self-contained operations included bunkhouses and kitchens. Entire families would travel from camp to camp for work. Note the wife, child, family dog, chicken, and ever-present workhorse in this scene. (COTA.)

Nathaniel Littlefield operated this portable sawmill near King Street. By the late 1800s, it was not necessary to build alongside a brook for waterpower. Steam boilers had come into use to power mill equipment. On this day in the late 1890s, a horse-drawn load of logs is arriving at the mill. Littlefield's operation closed when the boiler exploded in 1907. (SHS.)

A Stoddard Lumber Company crew is sluicing logs through a narrow waterway on the way to the mills on the North Branch River near the outlet of Island Pond. Samuel Robb opened his family's first sawmill in 1785. The family remained in the business, as four of his sons and two of his grandsons became millers. Grandson Christopher Robb opened his new pail shop on March 24, 1853. By 1870, his growing enterprise produced 8,000 boxes of clothespins, 40,000 pails, and 100,000 board feet of dimension lumber annually. Robb purchased thousands of acres of abandoned Stoddard farmland so he could cut and replant trees to supply raw material for his three mills. In the 1880s, he incorporated his business as the Stoddard Lumber Company. This photograph was taken on June 8, 1887, by amateur photographer Charles Newell, a Boston banker who summered in Stoddard with his family for 30 years. (SHS.)

Christopher Robb acquired water rights along a wide stream and series of ponds north of Mill Village and installed this dam in 1870. He now controlled the flow of water to power his woodenware mills downstream. The dam also raised the water level along the stream and the ponds, creating the seven-mile-long Highland Lake. This allowed the company to float its logs down the long lake. (COTA.)

Two Stoddard Lumber Company employees work to unload a log sledge near the mill on a winter day around 1900. The company's boardinghouse is visible in the background. The village that developed around the mills also included a company store, several homes, storage barns, stables, and the South Stoddard Post Office. Charles Merrill, a former millhand who married the owner's daughter, managed the mills by this time. (SHS.)

Christopher Robb built a woodenware empire. This view shows an addition being built at the company's Cherry Valley mills. Robb purchased more than 6,000 acres of land in Stoddard and surrounding towns to control both forest and water rights. In addition to the original pails and dimension lumber, the firm produced screw knobs, rocking chairs, handles for hand tools, baseball bats, and a variety of other wooden products. (COTA.)

This mid-1890s photograph shows several teamsters prepared to haul lumber and other products from the Stoddard Lumber Company to the Manchester & Keene Railroad station at Hancock to be shipped by train. The company employed dozens of townspeople in the mills, in the forest, and as teamsters to transport finished products to market. (SHS.)

It was reported that the health of the Stoddard Lumber Company declined under the management of Charles Merrill following the death of his father-in-law, Christopher Robb, in 1894. Despite this downturn, it was a serious blow to the town when the mills at Cherry Valley burned in 1908. An old resident declared, "When the mills went, the heart of the town went." (COTA.)

Charles Merrill is pictured here at the door of the new woodenware factory he built in Mill Village after his Stoddard Lumber Company in Cherry Valley burned in 1908. The new mill was located near his wife's general store and his Island House hotel in the village. The new factory produced chair stock for the local chair industry. This mill also burned to the ground in 1915. (COTA.)

# *Four*

# PUBLIC PLACES

Three young girls perform recitation at the South Stoddard school during the first decade of the 1900s. By 1820, the growing agricultural community boasted 11 one-room schools for a population of 1,203 people. There were school buildings throughout the town so the students could walk to the nearest schoolhouse to attend class. This typical classroom contained desks, benches, and a woodstove with a long stovepipe to distribute the heat. (SHS.)

The District No. 5 schoolhouse was in the village of South Stoddard. In 1862, the teacher here registered 49 pupils and reported that there were not enough textbooks. She also reported that the school had plenty of ventilation due to cracks and crevices in the walls. This schoolhouse was removed due to highway construction, and a new one-room school was built in the neighborhood. (SHS.)

This new schoolhouse replaced the original smaller school in South Stoddard. This view, taken in about 1912, shows a typical one-room school of the period with one entrance door for the girls and another for the boys. A flagpole stood in front, and there was an attic for storage. (SHS.)

The 21 students in the eight grades at the South Stoddard pose for a photograph with their teacher (right) around 1915. When the town's population decreased to 213 people in 1920, the South Stoddard school and the Stoddard Center school were the only schools remaining in operation. South Stoddard closed in the 1920s, and all students attended the Stoddard Center school. This building was removed in 1953. (COTA.)

The students in the Stoddard Center school summer term in 1864 pose for a class picture. The 21 students include only three boys. Most boys skipped the summer term because they were needed at home to work on the farm. The teacher, Sarah Blodgett, stands at the rear between the windows. This schoolhouse had been built 14 years earlier and would serve as a school for another 115 years. (SHS.)

By 1872, only seven of the original 11 schools survived. School No. 8, the north school, shown here, had few students to serve. The town population had decreased by more than 525 people since 1820. The total cost for schools that year was $999.95, or $6.89 for each of the 145 pupils. (SHS.)

The District No. 3 school (left) was known as the "Bog School" because of its location beside a mosquito-filled swamp. A scholar's strike was reported here in the late 1800s. The students claimed that the teacher was not giving proper instruction. A few students stayed home, and then all 17 pupils walked out. A new teacher was hired, and the students returned. (COTA.)

The 12 students in the Jefts neighborhood school pose in front of a US flag in the late 1800s. The pupils dressed in their finest clothes for the photographer. This was undoubtedly a rare sight when most students came to school in their work clothes and often attended barefoot. This school was no longer in use when it burned in the infamous Marlow-Stoddard forest fire of 1941. (SHS.)

The diminutive Leominster Corner school had been closed for years when this photograph was taken in the early 1900s. This neighborhood was settled by a group of families from Leominster, Massachusetts. Most of those families had moved on by 1900, and the schoolhouse was one of the last buildings to survive in the neighborhood. (COTA.)

Students at the Stoddard Center school were playing a ball game at recess when this photograph was taken around the 1950s. The school was built in 1850 to replace the previous school that burned. This one-room school served the town for 129 years and became one of the last operating one-room schoolhouses in the state. It was still being used as the town's only school when it burned in 1979. (COTA.)

The Stoddard Congregational Church was built in 1836 following the passage of the New Hampshire Toleration Act, which essentially separated church and state. Church organizations began to leave the old meetinghouses and construct their own church buildings. This photograph of the Gothic Revival structure was taken in about 1960. Sunday services are still held in the building. (COTA.)

Construction of the Stoddard Town Hall was completed in 1868. The large town meetinghouse had been sold in the 1840s, and the town government offices had been moved to a former church on this site. When the back wall of that building collapsed, this new town hall was built. It still serves as the town hall today. (SHS.)

The town post office was the social center of the community. It was important to pick up one's mail in the days before telephones and e-mail, and it was known that friends and neighbors would be at the office to pick up their mail. This group gathered on the porch of Frank Reed's general store and post office in 1902. The store closed within a couple of years, and the post office moved to another store almost one mile away. (SHS.)

The South Stoddard Post Office was located here (left) adjacent to the South Stoddard Glass Manufacturing Company boardinghouse in the mid- to late 1800s. The post office was established in 1842. It was discontinued briefly in the late 1840s but reestablished in 1850, after which it operated for 60 years without interruption. The office operated in at least three locations during those years. (COTA.)

The South Stoddard Post Office was located here in the James Stevens house from 1908 until the office was discontinued in 1910. Note the post office sign above the door. The office was moved to this location after the post office located in the village of Cherry Valley was destroyed by fire in 1908. The Stevens house was torn down in 1929 or 1930. (COTA.)

The Stoddard Fire Station was built in 1940 at a cost of $2,000. The need for a formal fire department and firefighting equipment became obvious the following year when the largest forest fire in New Hampshire history, the Marlow-Stoddard fire, burned several homes and 42 percent of the town's land area. The department featured its fire trucks in this c. 1950 photograph. (COTA.)

The original forest fire lookout tower on Pitcher Mountain was constructed in 1915. It consisted of a wood platform supported by four poles with a small shelter on top for the watchman. At 2,153 feet in elevation, Pitcher Mountain is the third-highest peak in Cheshire County, providing distant views in all directions to aid the watchman as he searched for forest fires. (COTA.)

The original lookout tower on Pitcher Mountain was replaced in 1925 by a steel tower with an enclosed observation room (left). A small cabin was constructed near the base of the tower so that the watchman could spend the night on the mountain or escape from severe weather on the mountaintop. A new steel tower was built after the Marlow-Stoddard forest fire destroyed the observation room of the lookout tower in 1941 (below). A new cabin was also built by members of the World War II Conscientious Objectors Camp located in Warner, New Hampshire. The tower can be accessed via a short trail from a parking area and has become a tourist attraction. (Left, COTA; below, HSCC.)

The Stoddard Historical and Village Improvement Society was formed at the Old Home Day celebration in 1907 with the express goal of purchasing the old Nathan Morse shoe shop in Stoddard Center village to be used as a town library and historical museum. The building opened in 1915, and the society purchased the property in 1917. This photograph of the museum was taken about 20 years later. (HSCC.)

The Stoddard Historical Society was inactive for about 15 years beginning in 1949, during which time the society building fell into disrepair. The museum was fully renovated in 1965 and reopened to the public. The society purchased the old hearse house and historical hearses from the town for $1 in 1969 and moved them to the historical society property that year. This photograph was taken in 2006. (SHS.)

A private library association was established in 1799. It would be 150 years before the town had a dedicated library building. The town established a public library in 1892. Louise E. Davis, librarian from 1926 to 1946, bequeathed $5,000 to the town for the construction of a library building. The building was completed in 1949, and this photograph was taken the following year. (SHS.)

Charles Peirce photographed the Old Stoddard Cemetery on Dow Hill during the winter of 1926. John Tenney deeded this piece of land to the town in the late 1700s to be used for a meetinghouse and burial ground. This is the oldest formal cemetery in town; the earliest gravestone bears the date December 25, 1792. The most recent of the 220 gravestones is dated 1907. (HSCC.)

## *Five*

# Highways and Byways

This road, known as the "county road," was laid out by order of the county government in the 1830s. The road was constructed from Keene through the town of Stoddard. This image was taken near the village of South Stoddard; the South Stoddard schoolhouse is visible in the distance. A large portion of the county road was bypassed upon the construction of the Keene and Concord Road in 1930. (COTA.)

This double-arch stone bridge was constructed on Antrim Road in the 1850s to replace previous wooden bridges that had been washed away by flooding. The sensible replacement has now stood for more than 150 years, but the selectmen who ordered its construction were never reelected because the townspeople believed the expense was too extravagant. The dry-laid stone arch bridge has been recognized as a civil engineering landmark. (HSCC.)

The stone arch bridge on the North Branch River was damaged by flooding in the 1990s. Although the bridge has long been bypassed by traffic on state highway Route 9, its importance as a historical landmark combined with strong encouragement from the townspeople convinced the New Hampshire Department of Transportation to repair the structure using the original dry-laid construction methods used when it was built in the 1850s. (HSCC.)

Mill Village was the site of the first sawmill and gristmill in the town. When this photograph was taken a century later, the village was still a busy manufacturing center. Two glass factories had located in the village in recent decades. Some residences, a retail store, and a tannery are visible in this image. (SHS.)

In the days before automobiles, a variety of transportation methods were employed to move from one place to another. These included traveling by foot, horseback, and a wide variety of wheeled vehicles. This industrious fellow rode through the center village on the back of a pair of gentle oxen owned by his employer, Henry Spaulding. (SHS.)

No rail lines were ever constructed through Stoddard, but several stagecoach lines traversed the community in the 19th and 20th centuries. A small open stage and a delivery wagon met at the South Stoddard Post Office on this day in about 1900. Items being delivered from one town to another were transferred at this stop, and then the two vehicles headed off in different directions. (HSCC.)

Stoddard was never the final destination of any of the stage lines, but the coaches stopped at stores, hotels, and post offices as they rolled through the town. Many summer visitors traveled by stage to come to the community in the early 1900s. This stage stopped by the Cherry Valley House and South Stoddard Post Office on this day in 1906. (SHS.)

Stage driver Horace Tuttle guides his three-horse coach along the main street of Stoddard Center on September 9, 1895. The Alstead to Hancock stage, known as the Forest Line, traveled daily between those two towns. The stage ride could be dusty and bumpy on the gravel roads but was preferable to walking the 27-mile route from Alstead to Hancock. (SHS.)

A fully loaded stage drawn by a three-horse hitch stops at Messenger's Store and the Stoddard Post Office on September 25, 1909. Post offices were a regular stop, as the mail arrived on the stage. The southbound coach arrived in town at 12:00 p.m., and the northbound coach pulled in at 2:30 p.m. The last horse-drawn stagecoach operated in 1915. (SHS.)

Three carloads of visitors arrive at the Island House Hotel in Mill Village a few years prior to the hotel's destruction by fire in 1917. The introduction of the automobile allowed for ease of travel and forced improvement to country roads throughout the state. Visitors such as these came to hike, swim, boat, fish, hunt, and enjoy other outdoor activities. (COTA.)

From left to right, Wynn Mountford, John Mountford, and John Hale work to clear Center Pond Road on March 25, 1916. Roads were cleared by hand to make a passage wide enough for a horse to pass through. Roads might eventually be widened enough for a wagon to travel. The town paid residents a few cents per hour to clear the town roads near their homes. (COTA.)

This early-1900s postcard is entitled "A Drive in Stoddard." The population of the town had been declining precipitously for almost a century when this photograph was taken. Consequently, postcards of improved gravel roads such as this view of Forest Road leading away from Mill Village were distributed to encourage nonresidents to visit the picturesque community. (COTA.)

High snowbanks line Forest Road in January 1957. Snowplows were used to clear the state highway by this time, but winter travel could still be treacherous. The town's only school, the one-room 1850 schoolhouse visible on the left, housed students in grades one through eight. Students were bussed to larger communities to attend high school. (SHS.)

This 1957 photograph of Forest Road (New Hampshire Route 123) near the top of Pitcher Mountain illustrates the nature of life in the winter in southwest New Hampshire in the mid-20th century. Snowstorms were frequent, and the snow was often measured by the foot. Stoddard had the reputation of being among the coldest and snowiest towns in the region. Several feet of windswept snow could fill the highway on the mountain. Heavy equipment was required to clear the two-lane road. The summit of the mountain is visible at the far right. Stoddard is physically one of the largest towns in the region. Consequently, as the population and the number of farms increased in the early 1800s, the number of town roads also increased rapidly. The town owned and maintained dozens of miles of roads by the mid-1800s. When the farms were abandoned and the population declined in the late 19th and 20th centuries, however, most of those town roads were discontinued, and the town owns and maintains only a small number of public roads today. (SHS.)

# *Six*

# Events of Man and Nature

Stoddard's "parade of horribles" lines up outside the general store in the town center in about 1868. Horribles parades, popular across New England in the late 19th century, consisted of townspeople dressed in grotesque and comic uniforms. These comical marching groups accompanied by brass and percussion instruments were a highlight of Fourth of July parades. Stoddard's horribles group was organized by Sarah Blodgett in the late 1860s. (HSCC.)

This picnic at Charles Merrill's Trixy Cottage on Island Pond was a highlight of Stoddard's participation in the first New Hampshire Old Home Day celebration in 1899. Gov. Henry Rollins inaugurated the statewide Old Home Day event to encourage former residents to return to visit their "old home" towns. Rollins envisioned this as a way to generate interest in the state and especially those towns with dwindling populations. (SHS.)

Old Home Day remained a popular event when this photograph was taken in the center village in August 1915. Townspeople and former residents gathered to enjoy picnics, speeches, lectures, church services, sporting events, and especially to visit with old friends. The tradition continues in towns across the state to the present day. (COTA.)

The highlight of the 1917 Old Home Day celebration was the dedication of the soldiers' monument on the lawn of the Stoddard Congregational Church, seen above. In the early days of August 1862, thirteen Stoddard men enlisted in New Hampshire's 14th Regiment of volunteer infantry. Approximately 120 men, more than 10 percent of the town's population, served in the Civil War. One of those men, 20-year-old James Hunt, enlisted with the group that joined the 14th Regiment in 1862. Fifty-five years later, James Hunt donated the soldiers' monument to the town. The inscription reads, "In memory of the men of Stoddard, who served as soldiers and sailors in all American wars." To the right, Selectman Cummings McClure (left) and donor James Hunt (right) pose with the recently unveiled monument. (Above, COTA; right, HSCC.)

Old Home Day

1836 1936

100 Years Old

August Eighteenth

1936

Stoddard, - - New Hampshire

Come Back Again!

Settled 1768 Incorporated Nov. 4, 1774

The town celebrated the 100th anniversary of the Stoddard Congregational Church building on Old Home Day 1936. This program proclaimed, "Come Back to the Old Home!" The offerings of the day included a picnic basket lunch, band concert, church service, and dance at the town hall (left). The 1936 event was the last Old Home Day celebrated until the tradition was revived for the town's bicentennial anniversary in 1974. The celebration has been held most years since that date and now runs for multiple days during the month of July. In recent years, the event has included a parade, chicken barbecue, children's activities, fairs and sales, and nature and historic hikes in the community. The photograph below shows one of the town's fire trucks in the 1989 Old Home Days parade. (Both, COTA.)

Baseball became very popular across the country in the late 1800s. The small town of Stoddard fielded its own team to play teams from neighboring towns. On September 14, 1895, the Stoddard nine climbed into this cart and traveled over Pitcher Mountain to neighboring Marlow to play the Marlow Hayseeds. (SHS.)

The town baseball team had its own uniforms by 1915, when this photograph was taken. The team did not survive many more years, partly because the availability of players decreased along with the population itself. Stoddard natives Bill Lane (first row, left) and Laurence "Pappy" Holmes (first row, center) remained in Stoddard for many decades after their playing careers ended. (COTA.)

Floods have often impacted the town because of rising water in the many streams and lakes in the community. Seven-mile-long Highland Lake overflowed the dam and dike at its southern end during the flood of 1927, seen here. One man is carrying another through the flood waters, and a third is wading along behind. This image was taken in Mill Village in front of Stewart's Store. (COTA.)

A recent snowstorm had visited Mill Village when this photograph was taken in February 1960. There were no snow blowers available, and someone had used the shovel visible in the pathway to clear a trail through the heavy snow from the door of the Thomas house to Shedd Hill Road. (COTA.)

Five former students pose during a reunion of the No. 2 school in the western part of town. These "girls" had attended the small one-room schoolhouse many years earlier. Reunions were held at the school regularly during the first half of the 20th century until the school building was destroyed by the Marlow-Stoddard forest fire of 1941. (SHS.)

This photograph captured the last regular church service at the Stoddard Congregational Church in September 1935. Church services were suspended after 100 years. The population was at its low point, and there were not enough members to continue. The church was subsequently revived and now offers weekly Sunday services once again. (SHS.)

The infamous Marlow-Stoddard forest fire of 1941 was the most damaging natural disaster in the town's history. The fire occurred in April. The hurricane of 1938 had left the forests choked with broken trees and branches. This material was very dry after months on the ground. Furthermore, April 1941 was exceptionally dry and warm. When an escaped spark at a sawmill in neighboring Marlow kindled a blaze, the resulting fire spread quickly from Marlow into the towns of Stoddard, Gilsum, and Washington. Approximately 2,000 volunteer firefighters converged to fight the flames. Some of them slept on the ground covered by buffalo robes (above). Rowboats were placed along the main street (below) to be used to pump water up the hill in Stoddard Center to fight the blaze as it approached the village. (Both, SHS.)

The Marlow-Stoddard forest fire raged across those two towns for three days in April 1941. One-half of the land area in Marlow was burned, and 42 percent of Stoddard went up in flames. The efforts of the firefighters had little effect in the forests, but they fought valiantly to save the village of Marlow and several individual homes in Stoddard. The fire jumped Highland Lake in Stoddard and continued to the east on the opposite side of the lake, burning several cottages in the process (above). A snowstorm arrived in the area on the fourth day and helped to squelch the flames. This was the largest forest fire in New Hampshire history, burning more than 20,000 acres. Several homes were destroyed, as seen below, but no one was killed or seriously injured in the famous blaze. (Both, COTA.)

School and church plays held on the town hall stage were long a tradition in the community. This early-1900s church youth play featured elaborate costumes and considerable evergreen foliage. Only two of the 17 participants in the play were boys. (COTA.)

The Stoddard Elementary School Christmas play in 1968 included all the school's 23 pupils. The one-room schoolhouse did not have a stage or adequate space for most public programs, so larger plays and public speaking events were held at the town hall by the mid-1960s. All graduates of the one-room school remember these public, often frightening, events performed before an audience of friends and family members. (COTA.)

An old-time country auction was held at the Davis House in Stoddard Center on July 6, 1946. The auctioneer stood at the front door selling antiques as they were removed from the house. Many auctions had been held in the town over the previous 100 years as families packed up a few belongings and moved to distant places in search of better financial opportunities. (SHS.)

Campers present a pageant in the 1910s at the end of their time at Camp Oahe, located on the shore of Granite Lake. Camp Oahe, operated by the family of Charles Eastman, a Santee Sioux, shared Native American traditions and activities. Oahe was one of numerous youth summer camps that operated in the town during the 20th century. (COTA.)

Stoddard celebrated its bicentennial anniversary in 1974. A large committee of volunteers worked for two years to prepare an appropriate celebration for a 200th birthday party. A wide variety of events were sponsored during the three-day celebration in July 1974. The activities kicked off with the New Hampshire Sweepstakes drawing at the town hall. Other events included dances, music events, house tours, children's activities, an art exhibit, church service, a barbecue, and historical tours and presentations. One of the highlights of the event was the long parade that wound down the main street. It was led by Gov. Meldrim Thomson Jr. (above, left). Stacy Eaton carried the sign with dates encompassing the town's first 50 years, as seen above. Stoddard's 1850s horse-drawn hearse was restored for the event and was featured in the parade, pictured below. (Both, SHS.)

*Seven*

# People of Stoddard

Moor Robb was a member of the large Robb family that settled near the village of South Stoddard in the 18th and early 19th centuries. Moor was a farmer who, like many early settlers, also undertook other economic endeavors to support his family. He opened a woodenware mill near his home and had an interest in the Box Tavern, built in the 1830s. (SHS.)

Christopher "Chris" Robb, son of Moor Robb, developed the largest business in the town's history. He opened a woodenware mill with his father in 1857. By the 1870s, Chris's lumber company was producing thousands of boxes of clothespins, tens of thousands of pails, and hundreds of thousands of board feet of dimension lumber annually. A village grew around the mills, and the company eventually owned 6,000 acres of woodland. (SHS.)

Joseph Foster launched another industry that had a major impact on the community. He opened a small glass factory in 1842. Five glass factories operated in the town over the next 31 years, producing millions of bottles and employing hundreds of residents. Joseph and his sons owned three of the Stoddard factories. The family remained in the glass business across the eastern half of the United States for five generations. (HSCC.)

Asa Davis Jr. (right) was born on the family homestead on the Keene Road where four generations of his family lived. He attended the Bog School, located about 50 feet from the family farmhouse. Asa married in 1843 and soon moved to a neighboring farm less than a mile away. Asa was known for his prize-winning steers and his large maple sugaring operation. Asa's wife, Sophronia Gould (below), was the daughter of Isaiah and Susan Gould. Isaiah was the first historian for the town, compiling the first town history in 1854. Asa and Sophronia raised a family of five children. (Both, HSCC.)

Rosabelle Davis, daughter of Asa and Sophronia, was born in Stoddard in 1845. This photograph was taken when she was about 20 years old. She was a schoolteacher and was described as well educated, refined, and highly esteemed in social circles. Rosabelle married Daniel Willard Rugg at the age of 27 in 1872 and moved to her husband's hometown. (HSCC.)

Twenty-year-old James H. Hunt enlisted in the 14th New Hampshire Regiment in 1862. He served in the military for three years until the war ended. His regiment traveled 15,000 miles and participated in at least 10 battles. Hunt donated the soldiers' monument on the Congregational church lawn to the town of Stoddard in August 1917. (HSCC.)

Conrad Webber immigrated to the United States from Switzerland in the 1850s. The Webber family moved to Stoddard before 1860 and took jobs weaving rattan onto bottles made at the Granite Glass Company in Mill Village. Eighteen-year-old Conrad Jr. (below) enlisted in a New Hampshire infantry regiment in September 1861 to fight in the Civil War. Conrad Sr. (right), aged 50, enlisted one year later. Conrad Jr. died after eating poisoned food in 1863. His father was captured at the Battle of Cedar Creek and died in a prisoner-of-war camp in Salisbury, North Carolina, in 1864. Both males of the Webber family died fighting to preserve their new home country. (Both, SHS.)

Stoddard native Jonathan Hale moved to Tennessee in the 1830s. He supported the Union cause when the war began and was forced from Tennessee by the Confederate authorities. Hale served as a military scout and spy for the Union army for more than four years. After clashing with the Ku Klux Klan after the war, he returned to Stoddard for the last years of his life. (COTA.)

Sgt. J. Langdon Reed served in the Civil War for almost three years. He was known as "the reckless fighter" for exposing himself to enemy fire during battle. Reed was called upon to act as guard for Confederate president Jefferson Davis soon after Davis was captured by Union authorities. Reed is shown here in his Grand Army of the Republic uniform. (COTA.)

Samuel Copeland (right) and his wife, Susan Richardson (below), were born one month apart in 1791. The couple settled in the north-central part of town in a neighborhood consisting of numerous Copeland family farms. Samuel was the 13th child of family patriarch Jacob Copeland. All the large Copeland family homes and barns are now gone. Cellar holes and stone walls mark the home sites in the reforested landscape. The former Copeland Cemetery is now deep in the woods, the family's remains having been removed to the New Stoddard Cemetery in the center village in the 1890s. Samuel and Susan posed for these portraits for posterity during the 1860s. (Both, SHS.)

Stoddard native Harvey Fisher studied to become a doctor and opened a medical office on the town common on Dow Hill in 1833. He soon moved his office and apothecary shop down the hill to the new town center. Children who attended the school across the street visited his shop after school hoping to get a licorice stick and to see the human skeleton hanging in the shop's closet. (SHS.)

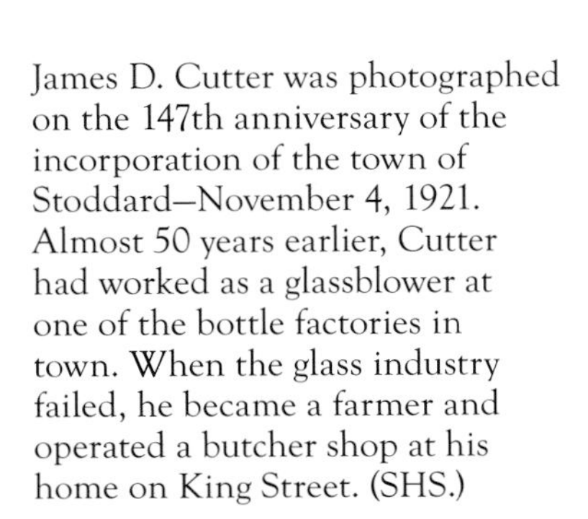

James D. Cutter was photographed on the 147th anniversary of the incorporation of the town of Stoddard—November 4, 1921. Almost 50 years earlier, Cutter had worked as a glassblower at one of the bottle factories in town. When the glass industry failed, he became a farmer and operated a butcher shop at his home on King Street. (SHS.)

Young farmer Harry Wilson was mowing his fields by hand when this photograph was taken in the early 1900s. Wilson was born in 1875 on the farmstead cleared by his great-grandfather in the 1790s. When most of the other farmers left Stoddard, Wilson became a cattle dealer, taking advantage of the abandoned pastureland. He eventually became the largest landowner in the town. (COTA.)

Happy and mischievous Pheroba Mountford posed for this photograph on her family's farm in August 1902, shortly before her fourth birthday. Mountford was educated at home by her parents and never had any formal schooling. She went on to become a poet, historian, curator, genealogist, and an inspiration to her family. (COTA.)

Frederick and Emily Reed were seated comfortably in the kitchen of their large farmhouse overlooking Center Pond when this photograph was taken in December 1901. Another year of hard work on the farm was coming to an end. A woodstove has replaced the fireplace behind it. Work gloves and socks are drying over the stove, and the family's beloved dog is stretched out in the warmth of the fire. Frederick and Emily's son George stands to the left of the stove. The main house at the Reed farm was built by Frederick's father, Jonas, in about 1839. He added a new structure to a smaller 18th-century house in which this kitchen was located. In 1901, this farm was one of only two remaining in a neighborhood around the pond that had been the site of 16 farms a half-century earlier. The Reeds were the last family to farm this land; the historic house was then purchased for use as a summer home. (SHS.)

Sarah Esther Minard hiked to the summit of Stacy Hill for a picnic in July 1913. Widow Sarah undertook this demanding outing during her 71st year. Her husband, Edmund, had died almost 30 years earlier, but Sarah continued to live in and manage the family homestead in Mill Village. (COTA.)

South Stoddard farmer Joseph Sweeney did not need a horse when his ox could be employed to plow his fields and pull his buckboard. Sweeney is pictured here in the early 1900s at the South Stoddard Post Office at Cherry Valley. Sweeney lived in one of the oldest houses in town, an active farm not far from where this picture was taken. (COTA.)

Longtime Stoddard farmer Henry Spaulding turned to cattle droving as a livelihood around 1900, by which time most other farmers had left town searching for greener pastures elsewhere. Spaulding is shown here with his oxen and oxcart. The farmer would bring about 250 cattle to Stoddard to pasture for the summer and would sometimes pasture hundreds of horses and up to 100 sheep on his Stoddard farmlands. (COTA.)

Fred Jennings, shown here at the Island House in Mill Village, arrived in Stoddard from Devonshire, England, at the age of 18 in 1888 and remained the rest of his life. He worked in a lumber mill and drove a stagecoach before becoming the first watchman in the new fire tower on Pitcher Mountain in 1915. Jennings later became a real estate agent and served as supervisor of the town checklist. (SHS.)

Farmer Ed Shedd was known as Stoddard's frugal Yankee. Although he was not poor, Shedd seldom spent money he did not have to spend. He walked to the county seat of Keene, almost 20 miles from his home, to conduct business rather than pay for a ride on the stagecoach. He often carried large amounts of cash to deposit in one of the 16 banks in New Hampshire and Massachusetts where he had accounts. Because his worn, old clothes looked like those of a tramp, no one suspected he was carrying large sums of money. He once walked to California and back to see more of the country. Following an injury late in life, he lived at the Elliot Community Hospital in Keene for the last four years of his life, paying his hospital bill daily. When he died at the age of 85 in 1934, he left his entire estate, more than $50,000, to the hospital. (COTA.)

Irene Eastman, daughter of Santee Sioux Charles Eastman, worked as a camp counselor at her family's youth summer camp on Granite Lake in the early 20th century. She helped teach the female campers about Native American history and culture. An aspiring actress and opera singer, Irene died during the Spanish flu epidemic in 1918 and was buried on a hillside at the camp. (COTA.)

Lifelong resident William "Bill" Lane was photographed as a young boy in about 1905. Lane epitomized service to one's community. He served many terms as a selectman beginning at the age of 22. He served on numerous other town boards and committees and represented the community in the state legislature for three terms. He also served uninterrupted as town moderator from 1936 until his death in the 1970s. (COTA.)

Charles Peirce first arrived in town in 1893 to visit a school friend. In 1902, he bought an old farmhouse in Stoddard, and for more than 50 years, he served as town historian, making detailed historical maps of the town, collecting historical notes and papers, writing about the town's history, and taking hundreds of photographs of the declining farm community. His decades of work made this book possible. (SHS.)

Florence Barbour was a summer resident who enjoyed the town so much that she moved there permanently. Her parents bought an old house in Mill Village in 1908 for use as a summer home. Barbour taught for 22 years at the Perkins Institute for the Blind before retiring to the house in 1952. During her 15 years in residence, she became well known and beloved throughout the community. (SHS.)

Florence Brooks Aten posed for this photograph in the 1920s on the porch of her rustic cabin, known as Shinbone Shack. A rich heiress with homes in Europe and New York City, Aten became enthralled with a plot of wooded land in Stoddard. She planned to live in the cabin while an Adirondack-style lodge was built nearby as her permanent home. Her dream also included the development of a pond and garden landscape surrounding her new home. Aten was a strong, outspoken woman who generously supported nonprofit causes she believed in. On one occasion, she sponsored a contest to develop a new national anthem for the United States. Her paradise in the woods was nearing completion when the stock market crash of 1929 took away all her wealth. The lodge and landscaping were left unfinished, and the property was sold at auction. With her fortune gone, Aten spent her remaining years in a small house in a neighboring community. (HSCC.)

# *Eight*

# Homes in the Country

The Parker house is the oldest surviving home in Stoddard. Early settler Oliver Parker built his home in 1772. He was one of the town's wealthiest and most influential residents, serving as first town clerk, first town moderator, and first town selectman. However, he remained loyal to Britain when the Revolutionary War began and lost favor in the community. He left when the war ended, but his home has remained for 250 years. (HSCC.)

Joseph Wilson built his farmhouse on the shore of Granite Lake in 1797. Four generations of his family farmed the land overlooking the lake. This photograph was taken during the last half of the 19th century before the home was enlarged. The property sold in 1918 for use as a youth summer camp. The old farmhouse was used as a summer home for much of the 20th century. (COTA.)

Charles and Martha Swain purchased the abandoned Sargent farmhouse in 1937 and carefully restored the house. Martha later opened an antique shop there. Built in about 1780, the old farmhouse was later home to Henry G. Foster of the Foster glassblowing family. Boston resident Louis Cabot used it as a hunting lodge when he purchased extensive acreage in the region to indulge his passion for hunting and fishing. (COTA.)

John D. Copeland built this large farmhouse for his growing family in 1801. John's father, Jacob, purchased land to give each of his sons enough property for a farm, and the Copeland family populated an entire neighborhood. Deacon Timothy Hunt lived here for many years in the 19th century, and his seven children were born here. The house burned in April 1905. (SHS.)

Asa Davis and his family lived in this early-19th-century Cape Cod–style home. The bicycle of Luther Smith, a later resident, is shown here with a crate of blueberries attached to the handlebars. Smith transported the berries several miles to the stage stop to ship them to market. The house was last used by loggers who were harvesting timber in the forests that reclaimed the former farmland. (COTA.)

The Nathaniel Worcester house, built around 1815, is seen in the foreground of this winter view of the old town center. Worcester served as the town doctor, but as the barn suggests, he also farmed the land, as did almost everyone else in town in the 19th century. The Worcester house survived for more than 150 years, but all the farmhouses visible in the photograph are now gone. (COTA.)

On Monday, August 24, 1936, town historian Charles Peirce walked into the overgrown field across the road to take this photograph of his house. Two Civil War soldiers, Congregational minister Samuel Gerould and farmer James Scott, had lived in the house before Peirce. A two-day auction held at the house disbursed many of Peirce's historical collections when he died in 1963. (COTA.)

This image of the Willard Spaulding farmhouse, photographed in 1916, illustrates the fate of hundreds of Stoddard farms. Willard's son, John Spaulding, served in the US Cavalry during the Civil War. John's mother, two sisters, one brother, and a niece died of diphtheria while John was in the service. John left the sad site after the death of his father, and the home collapsed some decades later. (SHS.)

Jonas and Sally Richardson married in 1830 and began the construction of their home overlooking Center Pond. Jonas Hadley purchased the home in 1866 and continued the farm for many years. The house was near collapse when it was purchased by Richard Hopkins in 1931 for use as a summer home. The restored house is now used as a year-round residence. (COTA.)

These two homes in Mill Village were built in the mid-1800s as tenements for workers at the Granite Glass Works, located in the village. The glass company operated there from 1846 to 1862. The firm employed 38 people in 1850 and produced 500,000 bottles. The two tenements were later converted to summer cottages and are now used as year-round residences. (COTA.)

Stoddard Lumber Company employee Adolph Miller built the home with the wrap-around porch in Mill Village. The house to the right was the home of Calvin Curtice, part owner of the glass factory located nearby. The house was purchased as a summer residence by the Reverend Thomas Barbour in 1908. The home is still owned today by members of his family. (COTA.)

Boston shipbuilder George Lawley purchased the c. 1840 Pitcher family house in Stoddard Center for use as a summer home in 1891. When the house was sold again more than 50 years later, it was advertised as being fully furnished, with well water, an electric generating plant, 60 acres of land, located on the state highway, and all for the price of $2,200. (COTA.)

Henry Wilson built this farmhouse in 1895, replacing a former house on the property that had burned in 1875. Opera singer and recording star Lambert Murphy purchased the house in the 1920s. It was purchased by the daughter of band leader and composer John Philip Sousa in the 1950s and has remained in her family since that time. (COTA.)

Paschal Hodgman built this two-and-one-half-story home on Marlow Road. The house was moved one-third of a mile to this site in the 1800s. The ell and porches were added after the move. The home was last occupied by the James Whippie family. The house burned in the Marlow-Stoddard forest fire of 1941. (SHS.)

Frank and Hattie Harlow had this Sears home built for them in 1913. Frank moved here from Jefferson City, Missouri, with the dream of becoming a gentleman farmer. Harlow chose a plot of land between Stoddard Center and Mill Village that looked down on Highland Lake and toward Shedd Hill in the distance. The home has been an architectural landmark in the town since that time. (COTA.)

This typical Stoddard farmhouse was built by the Henry family prior to 1806. Dexter Ball bought the house on Queen Street in 1856. Dexter, his wife Hanna, and the couple's seven children moved here from the neighboring town of Washington. This was a small farm with few livestock and limited crops being grown. The Balls owned the house into the 1890s, but it was later abandoned, as shown by this image. When a photographer arrived to shoot this photograph in the early 1900s, the barn had collapsed, windows in the house were missing or boarded up, and an abandoned ox cart stood in the front yard. Less than 20 years later, the house had experienced the same fate as many Stoddard farms, collapsing in upon itself. The shattered timbers and toppled chimney fell into the small cellar and quickly disappeared under the heavy hand of the New Hampshire climate. (SHS.)

The Adirondack-style Lakefalls Lodge was constructed for Florence Brooks Aten in the 1920s. Aten was a wealthy Rochester, New York, philanthropist who found this quiet spot in the early 1900s and decided to convert it into her paradise in the forest. She had a cabin built where she could live during the construction of the lodge. Work on the house and development of the landscaping began. A sunken garden, stone dam, boathouse, electric powerhouse, and stone walls were installed around a small pond, on whose shore the lodge would stand. The stock market crash of 1929 ended her dreams, however, and she lost everything. The Stoddard property, including the unfinished lodge, was sold at auction. A developer bought the property and subdivided numerous building lots, but the dam was condemned, and the development stalled. The lodge stood unfinished and abandoned for decades, becoming a favorite haunt for local teenagers during the 1950s and 1960s. The current owners completed Aten's dream many years after she began her lodge and her personal paradise in the woods of Stoddard. (SHS.)

# *Nine*

# Tourism

Tourism began to impact the community in the late 1800s, when people had sufficient time and money to leave larger towns and cities to travel to the countryside to enjoy the peace and quiet provided by nature. Stoddard offered numerous lakes and ponds and thousands of acres of forest. Here, two summer campers enjoy their time on the shore of Highland Lake in the 1890s. (SHS.)

The first tourists were sportsmen who came to enjoy hunting and fishing. The photograph above, dated August 1876, may be the first image of tourists in the town. This group came to fish in ponds that would soon become part of Highland Lake. They paddled up the connected ponds and camped on the shoreline. Groups such as this typically stayed several days and reported catching hundreds of trout, "bushels" of pickerel, and as many as 1,600 hornpouts in a single evening. The fellows (below) found a secluded cottage where they could fish, play baseball, smoke, and drink a lot, judging by the many empty bottles. (Both, COTA.)

Some farmers began to rent rooms and serve meals to summer visitors. Levi P. Spalding opened Mapleview Cottage to guests in 1891. The former William Wilson farmhouse on the old town road to Gilsum must have seen enough traffic to be economically viable. The public house was short lived, however, and it was removed in 1921. (SHS.)

This group of mostly young campers poses at their camp at Hunt Rock at Center Pond. Their handmade lean-to and smoking campfire are visible behind them. This photograph was taken in the 1890s by Boston banker Charles H. Newell. An amateur photographer, he produced some of the most important 19th-century photographs of the town. (SHS.)

Mill owner Christopher Robb brought in a crew from Boston to build this steamboat. Named *Myra* after his daughter and only child, the boat was used to haul logs to his mills, but it was also used to transport family members and visitors on Highland Lake excursions. Robb's son-in-law Charles Merrill would soon employ the boat to transport paying visitors to his rustic camps on the lake. (SHS.)

Charles Merrill's tourist camps were accessible by boat, adding to the backwoods, rustic nature of a stay on Highland Lake in the early 1900s. Merrill catered to groups as well as individuals, welcoming business and social groups that wanted to enjoy the beauty and bounty of nature, for a price. (COTA.)

The *Myra* was leaving the dock at Charles Merrill's tourist camps on the day this image was taken in about 1900. The steamboat not only transported visitors to and from the camps but also offered excursions on the large body of water. This was a common sight on New Hampshire lakes from the 1890s to the 1910s, when many steamboats plied the cool waters of the state and tourists were enticed with the promise of a comfortable excursion on the wild lakes of the state. Several lakes in southwest New Hampshire offered steamboat rides for guests at large resort hotels built beside the water. Many of these hotels offered all the comforts of home but were enhanced by the beauty of the New Hampshire wilderness. The *Myra* operated for several years but eventually sank to the bottom near its dock at Mill Village. (SHS.)

Merrill's Camp on Highland Lake offered its guests a select menu in the dining room in the camp's lodge, seen above. The lodge was surrounded by rustic log cabins for the guests to sleep in. Merrill also had a pet bear (below) that was an added attraction for visitors. Merrill rescued the bear as a cub and kept it as a pet for years. Merrill's Stoddard Lumber Company owned thousands of acres around Highland Lake and Island Pond. In addition to offering rural accommodations to visitors, Merrill tried to make a profit from those lands by selling small building lots along the lake shore after the lumber mills burned. (Both, COTA.)

This row of sleeping cabins consisted of part of the weekly accommodations available for visitors at Charles Merrill's Highland Lake camps. Annual clambakes held near the lake shore attracted visitors to the camps. Merrill published a booklet extolling the wonders of nature in Stoddard, which he described as "the Switzerland of New Hampshire." (COTA.)

These intrepid hunters were probably guests at the Swan cottage on Highland Lake when this photograph was taken in 1907. Former farmlands slowly reverting to forest and old apple orchards nearby provided excellent habitat for game birds and deer. This group sporting shotguns in early fall may have been in search of pheasants and ruffed grouse. (COTA.)

Boating quickly became a popular pastime in the town. Nine lakes and ponds of more than 10 acres in size, and five larger than 75 acres, attracted visitors to the water. The rowboat on Highland Lake (seen above) was photographed near the entrance to Pickerel Creek in the 1890s. It provided two sets of oars to ease the chore of rowing on the often-windy lake. The early motorboat below, decorated with an American flag, offered ease of travel to the two nattily dressed passengers in the early 1900s. This couple would be amazed by the large number of motorboats navigating the lake today. (Above, SHS; below, COTA.)

When Charles Merrill began to sell building lots around Highland Lake, small summer cottages started to appear along the shore. At least 13 people crowded Winona Cottage when this photograph was taken in 1906. The cottage was located on the eastern shore of the lake more than one-half mile north of Mill Village. (COTA.)

This tiny cottage on Highland Lake was catching the summer sun when this photograph was taken in the 1920s. The inhabitants of this diminutive cabin clearly enjoyed water sports, including fishing, boating, and swimming. Most of the cottages on the east side of the lake were accessible by road at that time, while many on the west side of the lake were reached by boat. (COTA.)

The Davis family cottage was more substantial than many of the camps on Highland Lake at that time, based on this 1909 postcard. Edward and Louise Davis eventually purchased a year-round home in town, as did many other summer residents. Louise Davis served as town librarian for 20 years and bequeathed money to the town for the construction of the first dedicated public library building. (COTA.)

The Harlow cottage on Highland Lake was typical of many of the early camps. This true log cabin offered a rustic wrap-around porch to provide outdoor space to enjoy the lake and the land. The cabin was fitted in among the trees with little disturbance to the land. As the 20th century progressed, new cottages became larger and were often surrounded by large lawns sweeping down toward the shore. (COTA.)

Trixy Cottage on Island Pond is one of the oldest surviving cottages in the town. It was owned by mill owner Charles Merrill when he hosted the town's first Old Home Day at the cottage on August 30, 1899. Trixy Cottage later took in guests and still serves as a summer home today. (COTA.)

Tourism increased not only on the lakes but on the highways as well. The J.D. Reynolds Inn was located at the intersection of Routes 9 and 123 in South Stoddard to cater to automobile travelers in the early 1900s. Reynolds offered an information station and the Stoddard Inn Restaurant and sold gasoline and camping supplies. (COTA.)

Paul Savage's summer music camp opened on the shore of Granite Lake in 1906. Youth summer camps gained in popularity in the late 1800s, and Stoddard's numerous lakes and thick forests made the town an ideal location for these back-to-nature camps. Savage's music camp lasted only a few years, but the camp lodge was used by other summer camps into the 1970s. (COTA.)

Robert and Eva Hodgdon bought the former Paul Savage summer campsite in 1926 and reopened it as a horseback riding camp for girls. Named Spruceland Camps, the catalog listed 56 activities the girls could participate in. The last youth camp on the site, Camp Notre Dame, was purchased by Alfred and June Skidmore in 1962. They operated the camp until its end in the 1970s. (COTA.)

Camp Winnecomack opened on the shore of Granite Lake in about 1910. The Winnecomack catalog suggested that "the physical well-being of your daughter must receive its full need of attention. The condition is best secured by a summer spent in the open." The open-air sleeping arrangements—tents on wooden platforms—can be seen on the hill behind the house. (COTA.)

Charles and Elaine Eastman opened their summer camp on Granite Lake in 1915. Numerous youth camps operated in the community between 1900 and the 1970s. All these camps stressed healthy outdoor activities for their youthful campers. One Oahe summer camper is about to take flight in a swan dive from the ledges on the island in the lake. (HSCC.)

A group of Camp Oahe girls paddled to the island at Granite Lake to partake of a cookout around 1919. Charles and Elaine Eastman named their camp Oahe from the Dakota word meaning "Hill of Vision." Charles Eastman was of Santee Sioux descent, and his camp stressed Native American history and culture in its programming for campers. He was widely known as the author of Native American and nature books as well as a leader in the Boy Scout and back-to-nature movements. Charles and Elaine Eastman's children took active roles in the camp. They also employed Winnebago artist Angel Dietz to teach outdoor sketching and artistic handicrafts, especially Indian beadwork and basketry. At Camp Oahe, the Eastman family sought to change the general American view of Indian culture by focusing their camp education on Dakota Sioux culture. They promoted the camp as "America's Only Indian Camp for Pale-Faced Girls." The camp is visible in the distance behind the campers. (COTA.)

Four Oahe campers work on camp activities at their platform tent. The camper on the right appears to be writing a letter home. Youth summer camps originally had campers sleep in the open air or in tents on the ground. Platform tents such as this one came next, and then cabins were introduced for health and safety purposes. (HSCC.)

Kinacamps was located on Highland Lake in the 1920s and 1930s. The name came from the fact that campers developed their own program for the summer "akin" to their own nature and personality. It was billed as a Western riding camp where horseback riding and Western-style clothing were employed. The cost was $285 for the eight-week season. (SHS.)

This rustic outdoor camp on an island in Highland Lake was photographed in 1931. The campers made considerable effort to prepare for their escape to nature. The tent has been pitched, the woodstove installed, the firewood cut, and the table built. The people who developed this camp were undoubtedly hoping to escape from the hustle and bustle of daily life; even their pet dog accompanied them on their adventure. (COTA.)

Camp Merriewood was one of a long line of youth, adult, and family camps at the site of Charles Merrill's former tourist cabins on Highland Lake. This group of family visitors was photographed at the camp in 1944. Arriving at the camp by boat, visitors spent their vacation time at the lake to relax, get back to nature, and escape the heat and noise of summer in the city. (COTA.)

# *Ten*

# Picturesque Lakes and Forests

Stoddard has long been known for its lakes and ponds. The bodies of water were little more than an impediment to the early farmers who cultivated crops. However, mill owners made use of the water to turn their waterwheels, and for the last 140 years, the placid surface of these lakes has been enjoyed by nature lovers. These fellows are boating on Island Pond in about 1910. (COTA.)

Highland Lake is the largest water body in the town and one of the largest in the region at 679 acres in size. It was formed from three small ponds when a dam was installed in 1870 to produce power for industrial use. This view looking north toward the town of Washington was taken in the late 1800s. (HSCC.)

Trout Pond is one of the smallest ponds in town, located in the rocky and hilly northeast corner of the community. All the farms that surrounded the pond in the 1800s have now reverted to forest. The former farmland around the pond is now conserved as property of the Society for the Protection of New Hampshire Forests. (SHS.)

This early-1900s view shows Highland Lake at a time when loggers and outdoor enthusiasts shared the water. Fishermen and boaters had to be careful to avoid the log boom in the lake. These logs were destined for one of the mills at the south end of the lake. (COTA.)

Cold Spring Pond was dammed in the 1790s to power a gristmill at its outlet. Robert Burnett, president of the Joseph Burnett flavoring extract company of Boston, purchased several local farms and the land around the pond during the first decade of the 1900s. He built a large camp for his family, which is visible in the distance near the shore of the pond. (COTA.)

The "Wheeler Bungalow," known as Greyledge, is one of the most historic cottages on Highland Lake. The visitor who sent this postcard in 1911 wrote, "This is where I went the last of June to stay for two weeks, but the mosquitoes drove me home in less than a week." (COTA.)

Nicholas Carr owned a farm near the shore of Highland Lake in the 1850s. This camp, identified on this postcard as "Carr's Cottage," is located on Carr Island, offshore from the site of the family farm. The inhabitants of this now-historic cottage enjoy an unparalleled view of the lake. (COTA.)

By the 1930s, the population of the town had dwindled to 113 people. Although there was little employment available for those who remained, some enterprising citizens made some extra income by renting boats to visitors. Some of those rowboats at Highland Lake can be seen here at the public landing alongside Forest Road, now Route 123, in Mill Village. (COTA.)

This 1990s aerial view explains how Island Pond received its name. The pond was dammed for power and recreational purposes. Easy public access has made the large pond an oasis for bathers, boaters, and fishermen. The town's first Old Home Day celebration was held here. In recent years, skating parties, picnics, firework displays, and other public events have been held at the pond. (COTA.)

This 1910 view of the Jonas Reed farm illustrates its prominent prospect of Center Pond and the surrounding country. Once there were 16 farms in the surrounding neighborhood. Now, there are a few cottages along the pond's eastern shoreline, but a considerable amount of the land along the shore is protected through conservation easements. (COTA.)

Granite Lake was once known as Factory Pond, suggesting its importance to several mills that took advantage of the water flowing from it. Numerous youth summer camps operated at the lake during the first few decades of the 20th century, and many campers returned to the region after their days at camp. Today, this lake is surrounded by homes and cottages, but the water remains clear, cold, and clean. (COTA.)

This iconic view is from the summit of Pitcher Mountain, the third-highest peak in the region, looking toward Mount Monadnock, the tallest peak in the region. The fields and farm buildings of Pitcher Mountain Farm are in the foreground. The Pitcher Mountain fire lookout tower is located on the summit. The rolling hills beyond are a riot of color in foliage season. The hiking trail up Pitcher Mountain is one of the region's most popular and is heavily used. Scenes such as this are common in the town, and most will not change because of extensive commercial development. Approximately two-thirds of the property in the community has been preserved as open land by conservation organizations or through the employment of conservation easements by private landowners. Much of this natural space is available for public enjoyment and draws outdoor enthusiasts from near and far. (HSCC.)

This view of Shedd Hill was taken with Highland Lake in the foreground. The hill was still mostly clear after decades of agricultural use. It is completely reforested today. This view is typical of numerous scenic views in the town, combining the natural beauty of woods, water, and rolling hills. (SHS.)

This early-1900s view of Bacon Ledge illustrates the New Hampshire granite so prevalent throughout the community. Bacon Ledge was so named because the rock striations in the ledge appear like slabs of bacon on close inspection. This spot has been appreciated for its views of Island Pond and Highland Lake since at least 1849, when three men carved their names and the date in the rock. (COTA.)

Hedgehog Hill, visible in the distance, was given that name because it looked like a hedgehog to local residents. At 1,900 feet in elevation, it is visible from much of the surrounding countryside. The boy in the foreground is standing in an overgrown farm field. Solitary apple trees stand in each of the fields. (COTA.)

This view shows the beauty of both Highland Lake and the countryside surrounding the long body of water. Lovewell Mountain is visible in the distance beyond the north end of the lake. Boaters, fishermen, and bathers spend long hours enjoying the varied landscape along the shore, while hikers take advantage of conservation land within walking distance of the lake. (COTA.)

The massive Stoddard Rocks are known far and wide because of their size and their location on the top of the 1,640-foot-high Carter Hill. This grouping of several glacial erratic boulders was deposited on the hilltop thousands of years ago by a retreating glacier. One of the massive rocks is about 35 feet tall and is estimated to weigh 960,000 tons. The boulders were dubbed "Stoddard Rocks" by historian and mapmaker Charles Peirce when he first saw them in 1900. He included them on his 1902 map of the town, and they have been known by that name since that time. These c. 1900 images of some of the rocks show the Chapman Boulder (above) and a woman standing beside one of the huge rocks (below). (Above, HSCC; below, COTA.)

For 150 years, hikers have enjoyed the walk to Stoddard Rocks from Highland Lake. These boys visited the rocks in the early 1900s, when there was still a good view of the lake in the distance. Today, much of the view is obscured by trees, but the boulders still impress hikers. (COTA.)

Even the small Mill Village, created because of mills located on the streams at that location, is picturesque from the air. The village is nestled amongst woods and water. Eaton's General Store, at the center of the photograph, was the heart of the village for decades. The small, historic village is located at the southern end of Highland Lake, visible at the very top of the image. (COTA.)

The summit of Pitcher Mountain can be seen in the distance of this photograph taken from Marlow Road, now Route 123. The open fields of the Knowlton farm, now Pitcher Mountain Farm, offered clear views of the Pitcher summit and the surrounding countryside. The farm was the only commercial agricultural operation in the town for several decades from the mid-1900s into the 21st century. (COTA.)

This bucolic scene shows the Goodspeed family relaxing on the lawn of their home in Stoddard Center. Architect Howard Goodspeed moved his family to Stoddard, where they renovated and enlarged a 19th-century house in the village. He also designed the town's Davis Public Library. From left to right, Lillian, Balcom, and Howard enjoy the distant view of Morrison Hill and Mount Stoddard in this mid-20th century photograph. (HSCC.)

# Bibliography


Gould, Isaiah. *History of Stoddard, Cheshire County, N.H.* Marlborough, NH: W.L. Metcalf, 1897.

History Committee of the Stoddard Historical Society. *The History of the Town of Stoddard, New Hampshire*. Stoddard, NH: Stoddard Historical Society, 1974.

Monadnock Historical Societies Forum. *The Nursery of Liberty: Schools and Education in the Monadnock Region*. Keene, NH: Historical Society of Cheshire County, 2009.

———. *The Power of Water: The History of Water Powered Mills in the Monadnock Region*. Keene, NH: Historical Society of Cheshire County, 2012.

Rumrill, Alan. *Five Days in August: Stoddard, New Hampshire in the Civil War, 1861–1865*. Stoddard, NH: Alan Rumrill, 2014.

———. *This Silent Marble Weeps: The Cemeteries of Stoddard, New Hampshire*. Decorah, IA: Anundsen Publishing, 1990.

Thompson, Sheila Swett. *Perley, the True Story of a New Hampshire Hermit*. Keene, NH: Historical Society of Cheshire County, 2008.

Yankee Bottle Club History Committee. *Yankee Glass: A History of Glassmaking in New Hampshire, 1790–1886*. Keene, NH: Yankee Bottle Club, 1990.